COMETS

COMETS
blazing visitors from deep space

DAVID APPELL

The Reader's Digest Association, Inc.
Pleasantville, New York/Montreal

COMETS

READER'S DIGEST PROJECT STAFF

Editorial Director
David Schiff

Editor
Charles Flowers

Senior Designer
George McKeon

Production Technology Manager
Douglas A. Croll

Editorial Manager
Christine R. Guido

CONTRIBUTORS

Author
David Appell

Art Director
Eleanor Kostyk

Copy Editor
Nora Reichard

Technical Reviewer
Scott Seagroves

Picture Research
Carousel Research, Inc.

Photo Editors
Laurie Platt Winfrey
Mary Teresa Giancoli

Photo Researchers
Van Bucher
Christopher Deegan

READER'S DIGEST ILLUSTRATED REFERENCE BOOKS

Editor-in-Chief
Christopher Cavanaugh

Operations Manager
William J. Cassidy

Art Director
Joan Mazzeo

Copyright © 2001 The Reader's Digest Association, Inc.
Copyright © 2001 The Reader's Digest Association (Canada) Ltd.
Copyright © 2001 Reader's Digest Association Far East Ltd.
Philippine Copyright © 2001 Reader's Digest Association Far East Ltd.

All rights reserved. Unauthorized reproduction, in any manner, is prohibited.

Reader's Digest and the Pegasus logo are registered trademarks of
The Reader's Digest Association, Inc.

Printed in the United States of America.

To order additional copies of *Out There: Comets*, call 1-800-846-2100.

You can also visit us on the World Wide Web at: **www.readersdigest.com**

Library of Congress Cataloging in Publication Data

Appell, David, 1960–
    Comets: blazing visitors from deep space / David Appell.
      p. cm.—(Out there)
    ISBN 0-7621-0313-2
    1. Comets. I. Title. II. Series.
    QB721 .A48 2001
    523.6—dc21   00-062581

Address any comments about *Out There: Comets* to:
Reader's Digest
Editor-in-Chief, Illustrated Reference Books
Reader's Digest Road,
Pleasantville, NY 10570

"No man is so utterly dull and obtuse, with head so bent on earth, as never to lift himself up and rise with all his soul to the contemplation of the starry heavens, especially when some fresh wonder shows a beacon-light in the sky."

— Seneca,
Quaestiones Naturales VII,
"De Cometis"

COMETS

blazing visitors from deep space

A ONCE-IN-A-LIFETIME COMET

It was an astronomical event unlike anything ever seen, a once-in-a-lifetime comet that dazzled astronomers worldwide. When comet Shoemaker-Levy 9 slammed into Jupiter in the summer of 1994, all of Earth's telescopic eyes were turned to witness and record the tumultuous show. Astronomers around the globe coordinated their efforts through electronic mail, sharing their scientific results and their unbridled amazement. Amateur astronomers, too, trained their backyard telescopes on the giant planet, discerning the huge dark spots that appeared like pox on Jupiter's gaseous surface. Shoemaker-Levy 9 was cometary history in the making.

The comet had been discovered fourteen months earlier by three astronomers working with a 26-inch mirror telescope at the Palomar Observatory in

Giant plums of soot rise above the clouds of Jupiter as fragments of Comet Shoemaker-Levy 9 pummel the planet's surface.

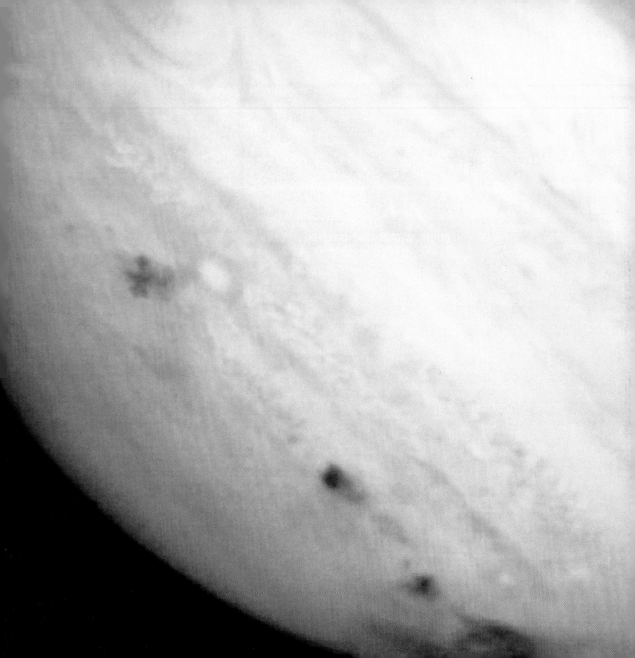

The dome for the 200-inch (500 centimeter) telescope at the Palomar Observatory is located atop Mount Palomar, northeast of San Diego, California. The inset shows the telescope.

California. The late Eugene Shoemaker and his wife, Carolyn Shoemaker, professional astronomers, and David Levy, a writer, lecturer, and amateur astronomer, were routinely surveying and photographing the sky on March 23, 1993. Two days later, Carolyn examined the film with a special stereomicroscope her husband had designed, looking for the characteristic stereoscopic effect caused by an object that had been moving in front of the background of stars. Unexpectedly, she found something that looked to her like a "squashed comet." Most comets consist of a large chunk of condensed ice, rock, and organic compounds several miles across, called the nucleus. As this nucleus nears the Sun, a bright, hazy shell called the coma develops around the nucleus. Closer still to the Sun, long tails of gases trail from the coma, pointing away from the Sun. These long tails, sometimes sweeping across much of the Earth's sky, create the seemingly magical auras that have thrilled Earth-bound observers since humans first looked up into the nighttime sky.

Curiously, the object found by Carolyn Shoemaker did not have a single coma and tail. Instead, it had a bar-shaped conglomeration of comae with a thin line of light at either end, with a composite tail stretching away. Her sighting was confirmed by a colleague at Kitt Peak National Observatory in Arizona whose more powerful tele-

The "string of pearls" that made up Comet Shoemaker-Levy 9, as seen by the Hubble Space Telescope in July 1994.

scope revealed astonishing news: There were at least five cometary nuclei, side by side, with comet material between them.

Alerted to the discovery by E-mail from the Central Bureau for Astronomical Telegrams in Cambridge, Massachusetts, astronomers around the world peered at the object for a closer look. Some reported a series of at least twenty separate nuclei, strung out "like pearls on a string"—Shoemaker-Levy 9 was in effect at least twenty separate comets. Within a month it was determined that the object was orbiting Jupiter, having passed very close to the planet eight months earlier. Astronomers had seen nothing like it before.

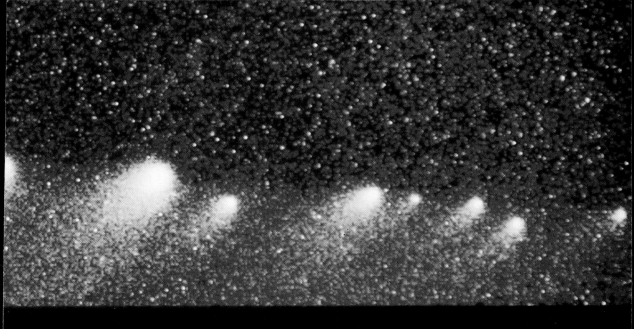

The story of the strange comet was now becoming clear. In 1992, Comet Shoemaker-Levy 9 had passed within 12,000 miles (20,000 kilometers) of Jupiter's cloud tops. As it made a turn around the massive planet, the gravitational tidal forces tore the comet apart. Further calculations indicated that the fragments would actually collide with Jupiter in July 1994.

This was major news. The collision of comets and asteroids with planets—especially Earth—has been a field of growing interest to scientists, especially since it was established in the 1980s that a comet or asteroid impact 65 million years ago led to the extinction of the dinosaurs. Comets rained upon the Earth during its formative period between 3.9 and 4.6 billion years ago, bringing a wealth of organic molecules—critical elements that allowed life to evolve. Scientists now had an opportunity to observe a planetary impact, not once, but twenty-one times, and they brought forth all the resources they could muster.

The Hubble Space Telescope was the most powerful eye available—its Wide Field and Planetary Camera was programmed to monitor the fragments as they plunged into Jupiter. Detailed calculations had predicted that the comet's fragments would strike, unfortunately, on Jupiter's night side, although the impact sites would rotate into Earth's view about an hour later. Meanwhile, the Galileo Space Probe, already on its way to a rendezvous with the planet, would be in position for its camera and other instruments to give a direct view of the impact sites. Observers using Palomar's 5-meter telescope, the Keck Observatory's 10-meter telescope in Hawaii, and many others throughout the world planned to observe the collisions in a variety of ways.

The collisions took place over six days, beginning July 16, 1994, in front of an expectant scientific world. The first fragment, labeled Nucleus A, slammed into Jupiter at 130,000 miles (210,000 kilometers)

Astronauts Story Musgrave and Jeffrey Hoffman repair the Hubble Space Telescope during a space shuttle mission in December 1993. During a series of five space walks, the crew installed corrective optics on the HST, replaced a camera, and completed other servicing tasks in time to view Comet Shoemaker-Levy 9's collision with Jupiter.

Sites where pieces of Comet Shoemaker-Levy 9 slammed into the Southern Hemisphere of Jupiter appear as brown spots in these images from the Hubble Space Telescope. Inset: The Fragment G impact site is seen as a bright spot at the center, surrounded by two ringlike features. The inner ring is 80 percent of the size of Earth.

per hour, creating plumes that rose almost 2,000 miles (3,200 kilometers) above the planet's clouds. One by one, as huge pieces slammed down, their planetary scars rotated into view, likened by one observer to "finger holes in a bowling ball." The largest spots could be seen by amateurs using 2.4-inch telescopes. Fragment G, which had a bright coma and was one of the largest, about 6 miles (10 kilometers) across, carried energy of 6 trillion tons of TNT—hundreds of times larger than the entire nuclear arsenal on Earth. Had it struck Earth instead, Fragment G would have left a crater about 100 miles (160 kilometers) across, about the same as was left by the comet or asteroid that caused the extinction of the dinosaurs. "Jupiter," said Levy, "is getting the stuffing knocked out of it."

Shoemaker-Levy 9's collision with Jupiter was one of the most notable events in astronomical history, and another example of the role comets play in science, in history, and in the human imagination. But what exactly are comets? Where do they come from? And might one someday slam into Earth? Sky watchers have been wondering since antiquity.

The beautiful colors of this comet, its coma, and its tails stand out against the star-speckled darkness of space in this illustration.

OUT THERE

HEAVENS BLAZE FORTH

Of the mélange of astronomical objects that make up the galaxy—stars and planets, moons and asteroids—comets are among the most mysterious. The stars rotate across the night sky on regular paths; our Moon rises and sets in a predictable fashion. But comets seem to appear at random, showing up unexpectedly, disappearing back into the void, each one different in brightness and the sweep of their tails. Small wonder the ancients saw them as mysterious harbingers of doom.

When they looked into their sky—a sky darker and richer than the light-polluted night skies of today—they occasionally saw bright, hazy objects that had not been there before. As the nights and weeks passed, the brightness increased, and the objects developed colorful, beautiful tails—wispy blends of blue, and sometimes green and red. These tails always pointed away from the Sun. Chinese astrologers, who were the first to record the sighting of a comet, in 1059 B.C., called these visitors "broom stars." Then the comets disappeared from view, which only added to their mystery.

Perhaps once or twice a decade, on average, and assuming a dark, rural sky, a comet will appear that is bright enough to be seen with the naked eye. "Great comets"—those that are the brightest, high

The Comet of 66 A.D., later named Halley's Comet, is depicted over Jerusalem in this 1666 rendering by an unknown artist.

A comet blazes over the manger in *Adoration of the Magi* painted by Giotto in 1286.

in the sky, in a position to be seen by a large number of people, even in cities—appear every decade or two. The Great March Comet of 1843 had a tail that was 180 million miles (290 million kilometers) long and ran from the horizon to the zenith, when seen from the latitude of Bombay, India.

Appearing unexpectedly, comets provoked fear among those who viewed them. The Incas regarded comets as a sign of anger from their Sun god, Inti. Death, usually of rulers and emperors, was expected with the appearance of the celestial torches. In *Julius Caesar*, Shakespeare's tragedy of political conflict and assassination, he writes

"*When beggars die there are no comets seen.*
The heavens themselves blaze forth the death of princes."

In actuality, a bright comet of 44 B.C. appeared several months after Julius Caesar's death, and was thought by some to be his soul going to heaven.

But anxiety about comets is not confined to the ancients. During the 1970 passage of Bennett's Comet, some Egyptians feared it was an Israeli secret weapon. And in 1997, thirty-nine members of the Heaven's Gate cult committed suicide upon the approach of Comet Hale-Bopp, apparently believing that an alien spaceship was following in its wake.

The most famous comet of all is probably Halley's Comet, named for English astronomer Edmund Halley (1656–1742; the name rhymes with "valley"). Halley had been trying to figure out the path of a comet he had seen in 1682 and, with the help of Isaac Newton and his new law of gravity, became convinced that it was the same comet seen in 1531 and again in 1607. All three comets traveled in an unusual direction around the Sun—opposite to the direction the planets orbit the Sun (what astronomers call retrograde). In 1705, Halley wrote, "Whence I would venture confidently to predict its return, namely in the year 1758."

Indeed, the comet was seen on Christmas Day 1758 by an amateur astronomer and farmer in Germany, and was seen again in 1835 and 1910. The interval between sightings has varied from seventy-five to seventy-nine years, with an average of seventy-six years. These small variations in Halley's orbits, called perturbations, are caused by the gravitational pull of planets, especially Jupiter.

The sighting of 1910 was especially anticipated, with much public interest around the world. But Halley's Comet was upstaged by an

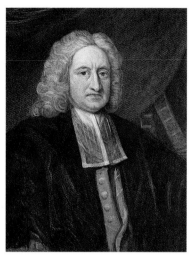

The English astronomer Edmund Halley (1656–1742) foresaw the return of the Great Comet of 1682, now known as Halley's Comet. It was the first prediction of a comet's return.

Top: Halley's Comet as it appeared above Egypt on May 25, 1910.

Inset: Consecutive images of Halley's Comet taken on May 15, 23, and 28, 1910, from Mount Wilson and the Palomar Observatories.

Halley's Comet in October 1986, photographed from New Zealand.

unexpected comet that could be seen even in daylight—the Great Daylight Comet of 1910. Comparatively, Halley's Comet, which appeared about two months later, was a visual dud. But predictions that the Earth would pass through Halley's gas tail (which had been incredibly long, at one point stretching over two-thirds of the distance from one horizon to the other) drove many unscientific Earthlings to despair. In Chicago, some without gas masks or the popular "comet pills" of the time stuffed rags in the cracks of their windows to keep out the toxic gas they mistakenly expected. One suicide was reported.

If humankind was more space literate for the comet's return in 1986, it was no less excited. Some ninty books about the comet were published, and it received wide media coverage during the time leading up to the expected sighting. Its appearance, though, was disappointing—the geometry of the comet, the Sun, and the Earth unfortunately combined to cause the tail to point mostly away from the Earth, where it was hidden by the coma. But five space probes were sent to investigate, one by the European Space Agency, two by the Soviets, and two by the Japanese. ESA's Giotto passed within the coma and obtained photographs of the nucleus, and a wealth of other scientific data. Halley's Comet will return in the year 2061.

It should be brighter than it was in 1986; but, as we'll discuss a little later, Halley's Comet, like all comets, is shrinking with each passage around the Sun.

THE SCIENCE OF COMETS

Only in the last fifty years or so have scientists succeeded in composing a complete theory of comets—what they are, where they come from, and what they can tell us about our solar system. We now know that comets are among the oldest, most primitive objects in the solar system—left over from the coalescence of a large cloud of dust and gas that began about 4.6 billion years ago. After the Sun formed in this collapse, and then the planets, smaller objects were left behind. Those composed of harder materials such as silicates we call asteroids, and those made mainly of ice we call comets (from the Greek, meaning long-haired).

Comets consist of a nucleus, a coma, and—when they are near the Sun—one or more tails. The nucleus is a relatively small chunk of various ices (water, methane, carbon dioxide, and other organic ices) and rock, typically a few miles in diameter. Fred Whipple, one of the pioneers of comet science, described the nucleus as a "dirty

The European Space Agency's Giotto Space Probe in 1986 passed within a few hundred miles (and inside the coma) of Halley's Comet and sent scientific data back to Earth.

The nucleus of Halley's Comet, as seen by the Giotto Space Probe.

PLASMA TAIL

DUST TAIL

NUCLEUS

COMA

A comet often has two tails streaming from the coma that surrounds its nucleus.

snowball." Halley's Comet, as seen by the Giotto Space Probe, has a nucleus about the size of Manhattan—about 5 by 9 miles (8 by 15 kilometers), with hills and craters and looking like a huge lump of charcoal.

But no comet nuclei can be observed from Earth because the nuclei are surrounded by a coma, a vast cloud of gas and dust. This coma is roughly spherical and comprises almost all of what we on Earth see of a comet. Giotto found that Halley's coma was about 80 percent water molecules, with most of the rest composed of carbon monoxide molecules.

The size of a comet's coma changes as the comet orbits the Sun. Solar energy tears gas molecules such as carbon monoxide away from the icy nucleus beyond the orbit of Jupiter, at about 5.2 A.U. (An A.U. is an astronomical unit, equal to the distance from the Sun to the Earth—about 93 million miles). When the comet is within 3.3 A.U. of the Sun, there is enough energy from the sun to convert its water ice directly to vapor (a process called sublimation), and the coma becomes prominent. Jets of gas and dust shoot out of active regions of the Swiss cheese–like nucleus; Giotto found seven jets on Halley's Comet, all spouting at the same time—a total of 3 tons per second of dust and 21 to 60 tons per second of gas.

Dust particles latch on to these water vapor molecules as they stream away at speeds of hundreds of miles per hour. (The appearance of a coma at 3.3 A.U. was, in fact, one of the early indications that the nucleus of a comet contained a great deal of frozen water.) A typical coma's radius is 60,000 miles (100,000 kilometers) or more, though some are much larger. The Great Comet of 1811 had a visible coma more than 1 million miles across—about the diameter of the Sun! There are estimated to be about 10^{13} (10 trillion) comets in our solar system, and perhaps as many as 10^{15} (a quadrillion), but with a total mass only a few times that of Earth. Individual comets have masses from 10^{10} to 10^{16} kilograms—a range from a few dozen Empire State Buildings up to a billionth the mass of the Earth.

For all that mass, comets are surprisingly sparse. Their nuclei are porous—the nucleus of Halley's Comet has been estimated to have a density similar to that of bread. Most of the coma is composed of particles the size of particles in smoke, though a few solid particles might be akin to pebbles or even small boulders.

But the poet in us is thrilled not by a comet's gravelly nucleus or its gassy coma, but by its long, spectacular tail. Comets develop tails as they near the Sun. As the nucleus's jets form the coma, these particles absorb and reflect the Sun's rays.

The coma and part of the tail of 1996's Comet Hyakutake, one of the brightest comets to appear in recent years is shown in this photograph.

The dust tail (green) and plasma, or gas tail (blue), stretch far behind a comet's coma as illustrated here.

Continual streams of particles emanate from the Sun in all directions. Some of them are photons, particles of light. These photons have momentum, so when they collide with another particle they exert a force. In the vacuum of space where there is no air or other resisting forces, the force of the particles' momentum can be quite substantial against the microscopic grains of dust. So when the Sun's light collides with the wispy particles in a coma, these particles are swept away, forming the comet's "dust tail." (This explains why a comet's tail always points away from the Sun.) At the same time, these photons reflect off the coma particles, illuminating the tail. Close observations of comet tails have revealed that comets actually have two tails. The dust tail, described above, is from 600,000 to 6 million miles (1 million to 10 million kilometers) long (the longest ever observed was 2 A.U.), and curves away as the comet swings through its orbit. The second tail, called an ion tail, or plasma tail, is bluish, points straight back, and is much longer, from 6 million to 60 million miles (10 million to 100 million kilometers). It consists of ions (charged atoms), and gets its brightness from fluorescence emanating from the ions, not from reflected sunlight. These cometary ions are created in the interactions between different phenomena: the sun's photons, the particles of the solar wind (the plasma, or ionized gas, that

Comet Ikeya-Seki displays its long tail on October 29, 1965.

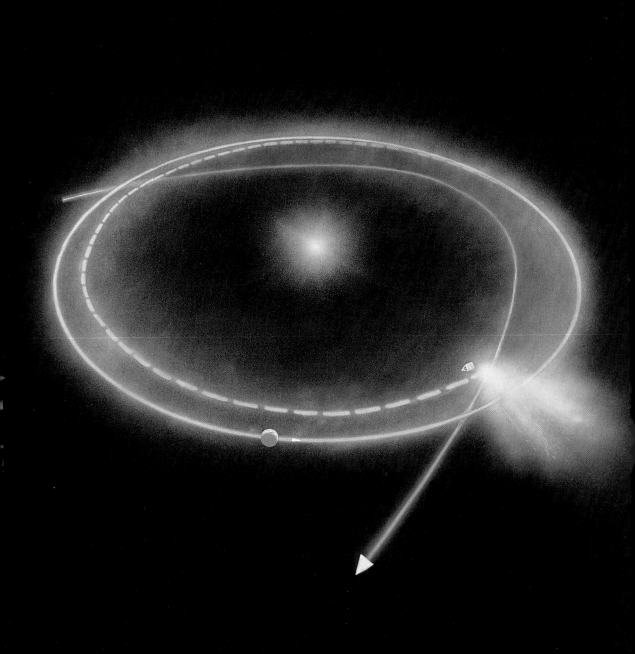

In this illustration of the encounter between the European Giotto Space Probe and Halley's Comet on March 13, 1986, the blue line shows the path of the comet as it travels around the Sun, the solid yellow line is the orbit of Earth, and the broken line is the path of Giotto.

is continually ejected into interplanetary space by the Sun), and the gas of the coma. Depending on a comet's nuclear composition, both types of tails may exist together in the same comet, or separately.

Comets typically travel an elliptical path around the Sun. (Shoemaker-Levy 9 was the first comet to be seen in orbit around a planet.) But their periods—the time required for one complete orbit—vary enormously. Halley's Comet takes about 76 years for each revolution. Comet Encke orbits among the inner planets, taking a mere 3.3 years for each cycle. Some comets take millions of years to complete one orbit, and some travel on parabolic or hyperbolic orbital paths and never return.

A comet's period can change over time. The shorter-period comets, which pass near the Sun relatively often, undergo repeated cycles of warming and cooling. As they warm, the jets of water vapor and gas carry away some of the nucleus's mass, which over time modifies the orbital dynamics of what remains. As well, comets frequently pass near planets, whose gravitational fields will slingshot the comet into a new orbit. In some instances, like Shoemaker-Levy 9, the resulting tidal force can even break apart the ice-rock matrix of the nucleus.

As the nuclear material leaves, it is gradually spread along the comet's orbit until this path is almost uniformly filled with tiny

OUT THERE

meteoroids—countless small, solid particles existing in interplanetary space. If our planet happens to pass through such a region of space, a meteor shower of "shooting stars" is the result. The oldest known meteor shower is the Lyrids, seen on April 21 of each year. Created by Comet Thatcher, the Lyrids have been observed for more than 2,600 years; Chinese records say "stars fell like the rain" in the shower of 687 B.C.. The notable Leonid shower of November 18th is linked to Comet Tempel-Tuttle. Halley's Comet, having passed the Sun more than a hundred times, creates its own shower as well, the eta Aquarid display

A Geminid fireball streaks through the night sky.

COMETS

Inset: Asteroid 243 Ida, about 31 miles (52 kilometers) in length, was photographed on August 28, 1993, by the Galileo Space Probe while on its way to Jupiter. This asteroid is a member of the Koronis family, orbiting the Sun between Mars and Jupiter. It is thought to have originated from a larger body that broke up. Its cratered surface shows that this is a relatively old asteroid.

that occurs around May 5 of each year.

Comets are, of course, named after their discoverers. There have been amateur astronomers, and a few professionals, who devote their lives to discovering new comets, standing on a chilly hilltop night after night. A name is their reward. If more than one astronomer independently discovers a comet, the name is hyphenated, up to a maximum of three. Comet Bakharev-Macfarlane-Krienke demonstrates the value of such a rule. Shoemaker-Levy 9 was the ninth periodic comet found by the Shoemaker-Levy team.

OUT THERE

THE OORT CLOUD

The fountainhead of comets is the Oort Cloud, named after Jan Oort, an astronomer who in 1950 proposed that many comets we see originate in a vast spherical cloud of comets surrounding the solar system. By Oort's time astronomers had determined that a great number of comets orbit the Sun, with the far point of the ellipse (called the aphelion) an incredible 50,000 to 100,000 A.U. from the Sun—more than one light-year. That's a thousand times farther from the Sun than the planet Pluto, and one-third the distance to our nearest star, Alpha Centauri.

Though there may be up to a quadrillion comets in the Oort Cloud, the cloud's volume is itself so immense that the average distance between its comets is about 1 A.U. You can't actually see the Oort Cloud—instead, it's a marvel of scientific inference.

Oort showed that passing stars could, through the long arm of their gravitational force, either pull comets out of the Oort Cloud or send them inward on their long path toward the Sun. Once near the planets, especially the behemoth Jupiter, they often are captured and become short-period comets (though sometimes they are instead ejected from the solar system), confined to planetary distances. In 1997, data from the European Space Agency's Hipparcos satellite

This artist's rendition shows the Oort Cloud, a spherical shell of space that extends from the Sun about one thousand times the distance to Pluto. The Oort Cloud is the main storehouse of comets in the solar system; it is estimated that it contains up to a quadrillion (10^{15}) comets.

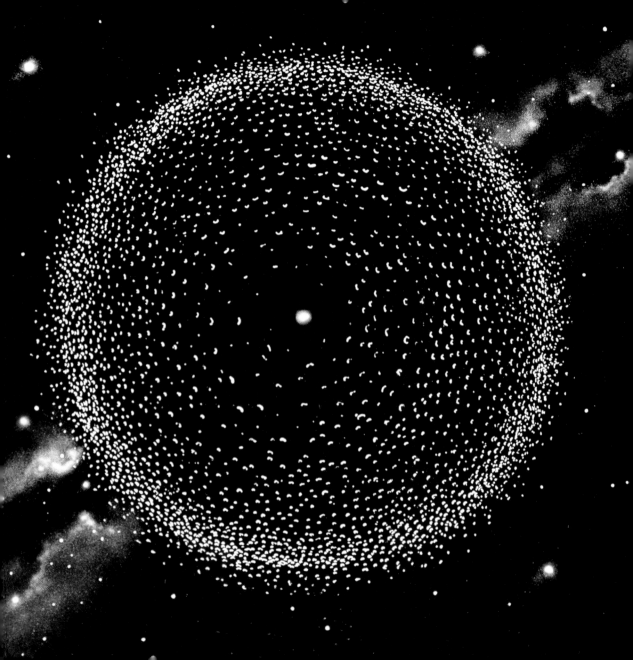

This is an artist's rendition of the Kuiper Belt, a vast population of small bodies orbiting the Sun in a thick ring or belt, extending from beyond the planet Neptune at 30 A.U. to about 50 A.U. There are at least 70,000 such bodies with diameters larger than 60 miles (100 kilometers). It is widely believed that the Kuiper Belt is the source of the short-period comets, acting as a reservoir for these bodies in the same way that the Oort Cloud acts as a reservoir for the long-period comets.

suggested that as many as eight stars will come within five light-years of the Sun in the next million years. Such stars could perturb the Oort Cloud, sending comets in all directions, including toward Earth. Most interestingly, data from Hipparcos suggests that the frequency of such disruptions is in accord with the frequency of mass extinctions on Earth.

Soon after Oort's proposal, Gerard Kuiper (pronounced Ki-per) postulated the existence of another grouping of comets, a flattened belt existing beyond the orbits of Neptune and Pluto—about 40 to 50 A.U. from the Sun. This inner cloud has come to be known as the "Kuiper Belt." Kuiper believed that comets in this region have their orbits disturbed by the large outer planets, sending them in all directions—some to become short- and medium-period, called Kuiper Belt objects. Kuiper's idea has gained credence as astronomers have detected several dozen extremely faint objects that move within the Kuiper Belt—comets that may visit us in the future.

Until recently, NASA was developing a robotic reconnaissance mission to explore the planet Pluto, its satellite Charon, and possibly an object in the Kuiper Belt. Called the Pluto-Kuiper Express, the mission is expected to launch sometime in the next two decades, with a flight time of eight to nine years. Many scientists suspect that Pluto, Neptune's moon may be related to Kuiper Belt objects, questions such a mission could help to answer.

THE ROLE OF IMPACTS

With the Oort Cloud sending comets in toward the Sun's planets, it's inevitable that one will slam into a planet occasionally, as Shoemaker-Levy 9 did. The most famous collision occurred on Earth, about 65 million years ago. By the early 1990s, paleontologists had determined that dinosaurs and about 60 percent of all other species then on the planet became extinct. Something drastic must have caused the extinction, because the dinosaurs had survived quite well for 150 million years.

The scientists focused on the so-called K/T boundary, the 65-million-year-old boundary between the Cretaceous and Tertiary periods. In 1980, physicist Luis Alvarez, his son Walter, and other colleagues announced the discovery of thousands of times more of the element iridium in a K/T boundary clay layer than is usual in the Earth's crust. The iridium signature has since been found in more than fifty locations around the world, and represents one of the most important scientific discoveries of the twentieth century.

This dinosaur is surely little aware of her fate in this artist's rendition of the comet or asteroid that struck Earth 65 million years ago, an event so colossal it's believed to have caused the extinction of many species, including the dinosaurs.

(continued on page 47)

RY B.C.	44 B.C.	1456	1577
that comets n of our own ere exhalations e ground and r atmosphere.	This comet, occurring several months after the murder of Julius Caesar, was said to be his soul ascending into heaven.	A comet appearing in the Mediterranean region caused so much fear that the Pope issued edicts requiring people to pray for salvation from the Devil, the Turk, and the comet.	Courtiers to Queen Elizabeth I of England tried to prevent her from looking at the comet and so tempt providence. Defiantly, the Queen strode to the window and gazed at it declaring, "The die is cast."

INFAMOUS COMETS

Imagine that you live in a time before electricity, before the invention of the telescope, before even the birth of Christ. In such a world there is no outdoor lighting to compete with the stars and planets, and the nighttime sky is alive with beauty and mystery. Your people have come to depend on its regularity—the rotation of the stars, the rising of the full moon, the positions of the constellations.

And then, as if created anew, an unrecognized star appears. But unlike the others, it grows brighter each night, spreading beyond a mere point of light to a smudge against the blackness. What is this visitor? As the nights go by, it brightens and lengthens, eventually stretching across a portion of the dark sky. An elder remembers another like it in the far past, and the disease that came the following year. Some believe it will grow until it swallows the darkness, and with it, the world.

It is mankind's nature to fear what we do not know, so it's small wonder that throughout history comets have been seen as harbingers of disaster and death. While science has now explained these luminous visitors, they still have the power to engender wonder and trepidation.

4TH CENT

Aristotle proposed were a phenomen air—that comets v of gases leaving th going into the uppe

The two opening pages show a detail from the Bayeux Tapestry, a 231-foot-long (70-meter) embroidery depicting the Norman Conquest of England in 1066. In the detail, a group of Englishmen point to the comet that appeared that year. Six centuries later the comet would be dubbed Halley's. The comet can be seen over the tower in the middle of the detail. At the right side of the detail, King Harold II of England is being told of the comet, which many believed to be a bad omen foreboding his defeat and death at the hands of William the Conquerer.

1973

A pamphlet by a group calling themselves Children of God explained that the comet heralded the end of the world on January 31, 1974.

1993

Carolyn Shoemaker, David Levy, and Eugene Shoemaker discovered the "string of pearls" comet, the 20 different pieces of Comet Shoemaker-Levy 9's nuclei that more than a year later slammed into Jupiter, providing the first human viewing of planetary impact.

1997

Encouraged by Art Bell, a popular nighttime radio talk show host who questioned whether a UFO was flying behind Comet Hale-Bopp, thirty-nine members of the Heaven's Gate quasi-religious group in San Diego committed suicide in the hopeful belief that the ship was coming to pick them up.

1577

Observations of a bright comet in 1577 led Tycho Brahe to the conclusion that, because the comet did not exhibit any parallax (a change in apparent position when viewed from different places on earth), it was at least four times farther away than the moon.

1665

The Great Comet of 1665 was visible in the sky when the black plague killed 90,000 people in London. Some thought the comet was to blame.

1664–66

Isaac Newton discovered his law of universal gravitation. Together with his development of the branch of mathematics known as calculus, this allowed scientists to predict the path of planets and comets through space.

The K/T boundary layer can be seen near Trinidad, Colorado.

(continued from page 44)

Where did the rare iridium come from? From a comet or asteroid that hit the Earth, said Alvarez and his colleagues. Scientists estimate its size to be about 6 miles (10 kilometers) across with a mass of a trillion tons (about 1,015 kilograms).

The effect of this enormous object smashing into the Earth was tremendous. It had the energy of about 100 million Hiroshima bombs, and launched 100 trillion tons (10^{17} kilograms) of vaporized dirt, rocks, and other material into the atmosphere. Earthquakes of magnitude 12 or 13 shook the surface, tsunamis half a mile (a kilometer) high roared through the seas, and vegetation the world over caught fire.

But that was the least of the problems. Ash, dust, and soot encased the Earth, blocking the Sun. For up to a year, day became indistinguishable from night. Temperatures plummeted, and photosynthesis, the essential process carried out by plants, ceased. Plant-eating species, which could not see in the dark anyway, soon died, and those that fed off them died soon after. The result was extinction of most of the world's species.

But where was the crater caused by the collision? Based on the size of the comet or asteroid, the crater created in the collision was estimated to be about 110 miles (180 kilometers) across and roughly 12 miles

An artist's rendition of the explosion as a comet smashes into Earth.

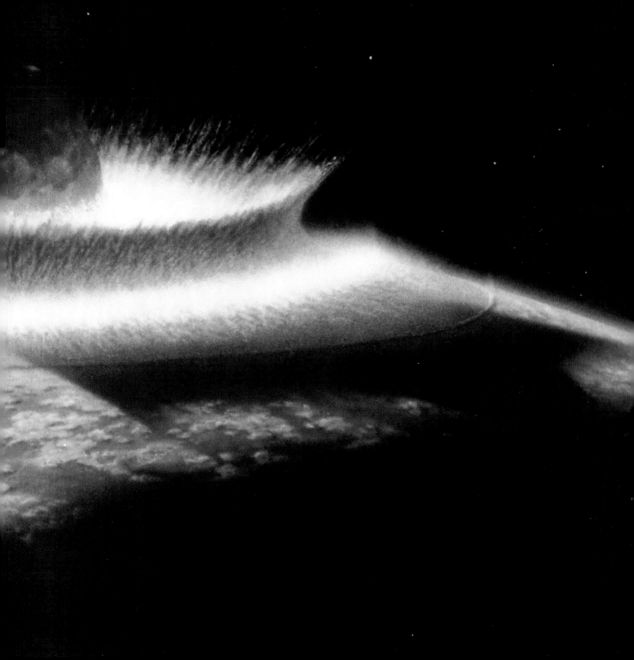

An illustration of the Chicxulub impact crater on the Yucatán Peninsula, Mexico, soon after its creation.
Inset: The location of the crater site at Chicxulub.

(20 kilometers) deep. By 1991, sleuthing geophysicists had found the missing hole—a 110-mile-wide crater lying underneath the surface rocks near the Mexican village of Puerto Chicxulub on the northern coast of the Yucatán peninsula.

Although drastic, this object smashed into the Earth eons ago. What if something similar hit the Earth in modern times?

It turns out something already has. In the Tunguska Event, a large extraterrestrial object described as "a piece broken off the Sun" exploded over central Siberia on June 30, 1908. Trees fell straight away from the blast for 30 miles (50 kilometers), the heat wave was felt in a village 37 miles (60 kilometers) away, and the sound of the explosion was heard 600 miles (1,000 kilometers) away. Scientists still aren't quite sure whether the object was a comet nucleus (the leading theory is a piece of Comet Encke that broke away) or a meteor. Had it struck a major city like St. Petersburg, a million people might have died.

Events like the Chicxulub comet, Tunguska, and Shoemaker-Levy 9 have made many on Earth realize that, viewed over millions of years, our planet might as well have a large bull's-eye inscribed on it—a

Trees were flattened by the Tunguska Event in Siberia, June 30, 1908.

fear exploited in the 1998 movie *Deep Impact*. Since 1700, the closest any comet has come to Earth was Comet Lexell in 1770, passing about 1.4 million miles (2.2 million kilometers) from Earth, or about six times farther away than the Moon.

But Earthlings are growing jittery. In 1998 a Harvard-Smithsonian Center for Astrophysics announcement that mile-wide asteroid 1997 XF11 would pass within 30,000 miles (50,000 kilometers) of Earth in the year 2028 commanded everyone's attention. A day later, new data and new calculations indicated that it would miss Earth by 600,000 miles (1 million kilometers)—still closer than any previously observed asteroid or comet.

Since then, astronomers have grown understandably wary of announcing the world's demise, and more scientific about it, too. The Torino scale, first created by MIT Professor Richard P. Binzel in 1995, is a "Richter scale" for categorizing Earth impact hazards for newly discovered comets and asteroids. Based on an object's trajectory and mass, the Torino scale (the name comes from a conference setting in Italy where it was presented) assigns it a number from 0 to 10, with 10 indicating a certain collision and 5 considered a "threatening event."

What would happen if a half-mile-wide (about 1-kilometer-wide) comet struck the ocean? In 1997, Sandia Laboratory scientist David

A scene from the 1998 movie *Deep Impact*, depicts a comet collision with Earth and the resulting dust cloud.

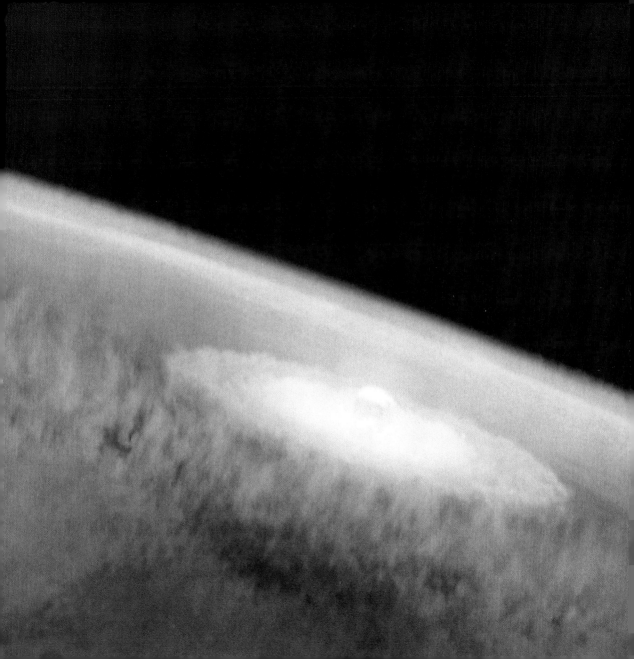

In this computer simulation of a comet impact on Earth, the comet has struck an ocean, throwing large amounts of high-pressure steam (green) into the atmosphere. The comet was assumed to be 1 kilometer wide, with an impact energy equivalent to ten times the explosive power of all nuclear weapons on Earth. For reference, the city silhouette is 2 miles (3 kilometers) wide.

Crawford carried out a computer simulation of such an event using a new teraflop supercomputer, capable of one trillion operations per second. Such a comet, about the size of the largest fragment of Shoemaker-Levy 9, would weigh about a billion tons (1,012 kilograms)—only 0.1 percent of the Chicxulub comet. Crawford assumed it would be traveling at 37 miles (60 kilometers) per second, and would enter the Earth's atmosphere at an angle of 45 degrees.

Using previously developed Sandia software called "bang and splat" code, Crawford predicted that the hypothetical comet would hit the ocean with an impact energy of 300 billion tons of TNT—ten times the nuclear explosive power of all nuclear weapons at the height of the cold war. It would smash into the ocean floor and vaporize, along with 120 cubic miles (500 cubic kilometers) of water. Six hundred miles (1,000 kilometers) from the impact site, the ocean waves would be 65 yards (60 meters) high. Low-lying areas such as Florida would be inundated.

It has been estimated that a comet or asteroid with this energy strikes Earth every three hundred thousand years—in other words, each century there is a 1-in-3,000 chance of such a collision. You'd probably play the lottery with those odds. In a cosmic sense, we already are.

RECENT COMETS

Comet West displays its dust and gas tails on March 9, 1976.

"Comets are like cats," David Levy once said. "They have tails, and they do precisely what they want." Over the course of history, comets have thrilled and frustrated observers and offered scientists a window into the early solar system.

Would that all comets lived up to the billing of Shoemaker-Levy 9. The biggest flop of all comets is no doubt 1973's Kohoutek, at least in the eyes of the public. The comet was expected to be one of the brightest in history, rivaling Venus, and with months of advance warning there was ample time for hype. But early predictions did not prove true—the comet's tail spanned only about 10 degrees and the only observers to get spectacular views were the Skylab astronauts.

Kohoutek fizzled because as a first-time visitor from the depths of the Oort Cloud it appeared deceptively bright. This brightness was caused by the vigorous vaporization of its virgin surface of volatile compounds. The public, expecting a truly "Great Comet," was left disappointed.

Astronomers, having learned a painful lesson, gave little advance warning for 1976's Comet West. West nearly was as bright as Venus, detectable even in daylight with telescopes and binoculars, with a bright, complex, striated tail. West's nucleus had split into at least four pieces that traveled together, shining brighter than other comets because more surface material was exposed to sunlight. Alas, Comet West will not return for about 558,000 years.

Comet Hyakutake—the Great Comet of 1996—did not disappoint. It was discovered early in the year by Yuji Hyakutake, a Japanese amateur astronomer. Calculations indicated that Hyakutake would pass within a relatively small 0.23 A.U. of the Sun and 0.10 A.U. (only 9.5 million miles or 15 million kilometers) of Earth, and that it had an intrinsic brightness comparable to Halley's Comet. Moreover, Hyakutake had last been in the inner solar system eight thousand years ago, so it had already burned off its virgin sheath. In March, the comet was high in the Northern Hemisphere sky and as bright as the brightest stars, with a tail spanning over 70 degrees. The Hubble Space Telescope revealed small fragments within the coma that appeared to have broken off the nucleus.

In the 1990s, cometary luck came in pairs, because after Hyakutake astronomers still anticipated Comet Hale-Bopp. One of the most notable comets in recent years, Hale-Bopp was discovered independently in July 1995 by Alan Hale and Thomas Bopp while it was still an incredible 7.2 A.U. away from the Sun (between the orbits of Jupiter and Saturn). That meant it was extraordinarily

This close-up shows the coma and tail of Comet Hyakutake, one of the brightest objects to appear in the sky in recent years.

Comet Hale-Bopp streaks over Tucson, Arizona, in March 1997.

bright—one thousand times brighter than Halley's Comet at the same distance—due to a surprisingly big coma. Comparisons to the Great Comet of 1811 were offered, and astronomers waited with excitement.

This time they were not disappointed. Hale-Bopp made its closest approach to Earth on March 22, 1997, at a distance of 1.3 A.U. The comet glowed as bright as the very brightest stars, visual even in the hardly dark skies of suburbia. The comet also proved to be a scientific bonanza, and its importance to cometary research was compared to the Halley encounter in 1986. One interesting discovery was that of a third tail on the comet, a thin one made of sodium atoms. Over five thousand images have been posted to NASA's Hale-Bopp Web site.

Though most comets are not visible to the naked eye like Hale-Bopp, we can now see comets with eyes unlike any available to our forebears. The Solar and Heliospheric Observatory (SOHO) satellite has found more than a hundred new comets, making it the most successful comet hunter in history. While its Large Angle and Spectrometric Coronagraph instrument watches for ejections of plasma from the Sun's surface, comets falling toward the Sun often show up as well. Most of these comets are what are known as "sungrazers," kamikaze comets that plunge into the Sun.

Researchers believe that many of these newly discovered sungrazers are fragments from the gradual breakup of a great comet, perhaps one the Greek astronomer Ephorus reported seeing split in two in 372 B.C.. Such a history of splitting can give clues to the internal strength and makeup of a comet's nucleus.

In February 1999, NASA's *Stardust* spacecraft blasted off from the Kennedy Space Center for an encounter with the periodic Comet Wild-2. The mission's goal is to intercept the comet in 2004, capture some of the comet's dust and debris, and return the samples to Earth for analysis in 2006. *Stardust* is the first cometary mission since Giotto's flyby of Comet

Above: This drawing shows the *Stardust* spacecraft, launched in 1999 to rendezvous with Comet Wild-2 in 2004. *Stardust* will gather dust spewed by the comet and return it to Earth for a detailed analysis.

Halley in 1986 and Comet 26P/Grigg-Skjellerup in 1992 and the first to return material to Earth.

Comet Wild-2 was discovered in 1978. It is a relatively new comet to the inner solar system, so it contains plenty of fresh material, leftover, primordial material like that from which the Sun and planets formed. By the time *Stardust* meets Wild-2, the comet will have passed around the Sun only five times.

Stardust will intercept Wild-2 well past the orbit of Mars, passing within 60 miles (100 kilometers) of the nucleus at 13,600 miles per hour, at which speed dust particles will become embedded into the craft's aerosol sample collectors. The craft will then return to Earth, the collectors settling by parachute in Utah.

In April 2000, scientists announced that comet tails might be quite a bit longer than previously suspected. This serendipitous discovery came from reanalyzing data captured by the *Ulysses* spacecraft in 1996, when it unexpectedly glided through ionized gases from the tail of Comet Hyakutake. The tail was found to extend more than 300 million miles (500 million kilometers), three times the distance from the Sun to the Earth.

Ulysses' discovery proves that comets can still surprise us. What lies in the future—a manned mission that lands on a comet nucleus? Will comets provide more information about the birth of our solar system or the origin of life on Earth? Or will the next surprise be another great comet whose tail sweeps across the sky, reminding us of the beauty of the heavens?

ABOUT THE AUTHOR

David Appell is a science writer and journalist; his writing on astronomy and astrophysics has appeared in *Popular Science, Discovery Channel Online, New Scientist, Current Science*, and other magazines. He holds a Ph.D. in physics from the State University of New York at Stony Brook.

COMETS

Picture Credits

Cover Joe Tucciarone/Science Photo Library/Photo Researchers **Back cover** Art Resource, NY **Title Page** John Chumack, John Bryan State Park Observatory, Yellowspring, Ohio/Photo Researchers **Gatefold** (opener) Erich Lessing/Art Resource (interior) (background) Christopher Short (1) The Granger Collection (2) Giraudon/Art Resource, NY (3) Culver Pictures (4) Scala/Art Resource (5) Art Resource, NY (6) The Granger Collection (7) The Granger Collection (8) Associated Press (9) Ted Melis/USGS (10) Associated Press/Las Vegas Sun **Endsheets** John Sanford/Starhome Observatory **5** David Nunuk/Science Photo Library/Photo Researchers **7** Space Telescope Science Institute/NASA/Science Photo Library/Photo Researchers **8** Bill W. Marsh/Photo Researchers **8** (inset) Lawrence Migdale/Photo Researchers **10** Space Telescope Science Institute/NASA/Photo Researchers **13** NASA/Science Photo Library/Photo Researchers **14** (both) Space Telescope Science Institute/NASA/Science Photo Library/Photo Researchers **16-17** Mehau Kulyk/Science Photo Library/Photo Researchers **19** Stanislaus de Lubienietski/ Historia Universalis Omulum Cometarum, Amsterdam, 1666/ Art Resource, NY **20** Scala/Art Resource, NY **22** The Granger Collection, New York **23** NASA/Photo Researchers (inset) Mount Wilson and Palomar Observatories/Smithsonian Library **24** Barney Magrath/Science Photo Library/Photo Researchers **26** European Space Agency/Science Photo Library/Photo Researchers **27** Comets & Comet Halley Astron Society of the Pacific/Photo Researchers **28** Christopher Short **31** Gordon Garradd/Science Photo Library/Photo Researchers **32-33** Mark Garlick/Science Photo Library/Photo Researchers **35** Science Photo Library/Photo Researchers **36** Julian Baum/Science Photo Library/Photo Researchers **38-39** John Sanford/Astrostock **39** NASA/Science Photo Library/Photo Researchers **41** Jon Lomberg/Science Photo Library/Photo Researchers **42** Christopher Short **45** Julian Baum/Science Photo Library/Photo Researchers **46** FranŸois Gohier/Photo Researchers **48-49** Don Davis/NASA AMES **50** (both) D. Van Ravenswaay/Science Photo Library/Photo Researchers **51** Novosti Press Agency/Science Photo Library/Photo Researchers **53** Paramount Pictures/Photofest **54** Sandia National Labs/Science Photo Library/Photo Researchers **56** Royer & Padilla/Science Photo Library/Photo Researchers **59** Rev. Ronald Royer/Science Photo Labrary/ Photo Researchers **60** Kent Wood/Photo Researchers **62-63** NASA/Science Source/Photo Researchers